LONDON BOROUGH OF ENFIELD
LIBRARY SERVICES

This book to be RETURNED on or before the latest date stamped unless a renewal has been obtained by personal call, post or telephone, quoting the above number and the date due for return.

OLYSLAGER AUTO LIBRARY

Tanks & Transport Vehicles World War 2

compiled by the OLYSLAGER ORGANISATION

edited by Bart H. Vanderveen

FREDERICK WARNE & Co Ltd
London and New York

THE OLYSLAGER AUTO LIBRARY

This book is one of a growing range of titles on major transport subjects.
Titles published so far include:

The Jeep
Half-Tracks
Scammel Vehicles
Fire-Fighting Vehicles
Earth-Moving Vehicles
Wreckers and Recovery Vehicles
Passenger Vehicles 1893–1940
Buses and Coaches from 1940
Fairground and Circus Transport

American Cars of the 1930s
American Cars of the 1940s
American Cars of the 1950s
American Trucks of the Early Thirties

British Cars of the Early Thirties
British Cars of the Late Thirties
British Cars of the Early Forties
British Cars of the Late Forties

Library of Congress Catalog Card No. 74-80617

ISBN 0 7232 1808 0

Filmset and printed in Great Britain
by BAS Printers Limited, Wallop, Hampshire

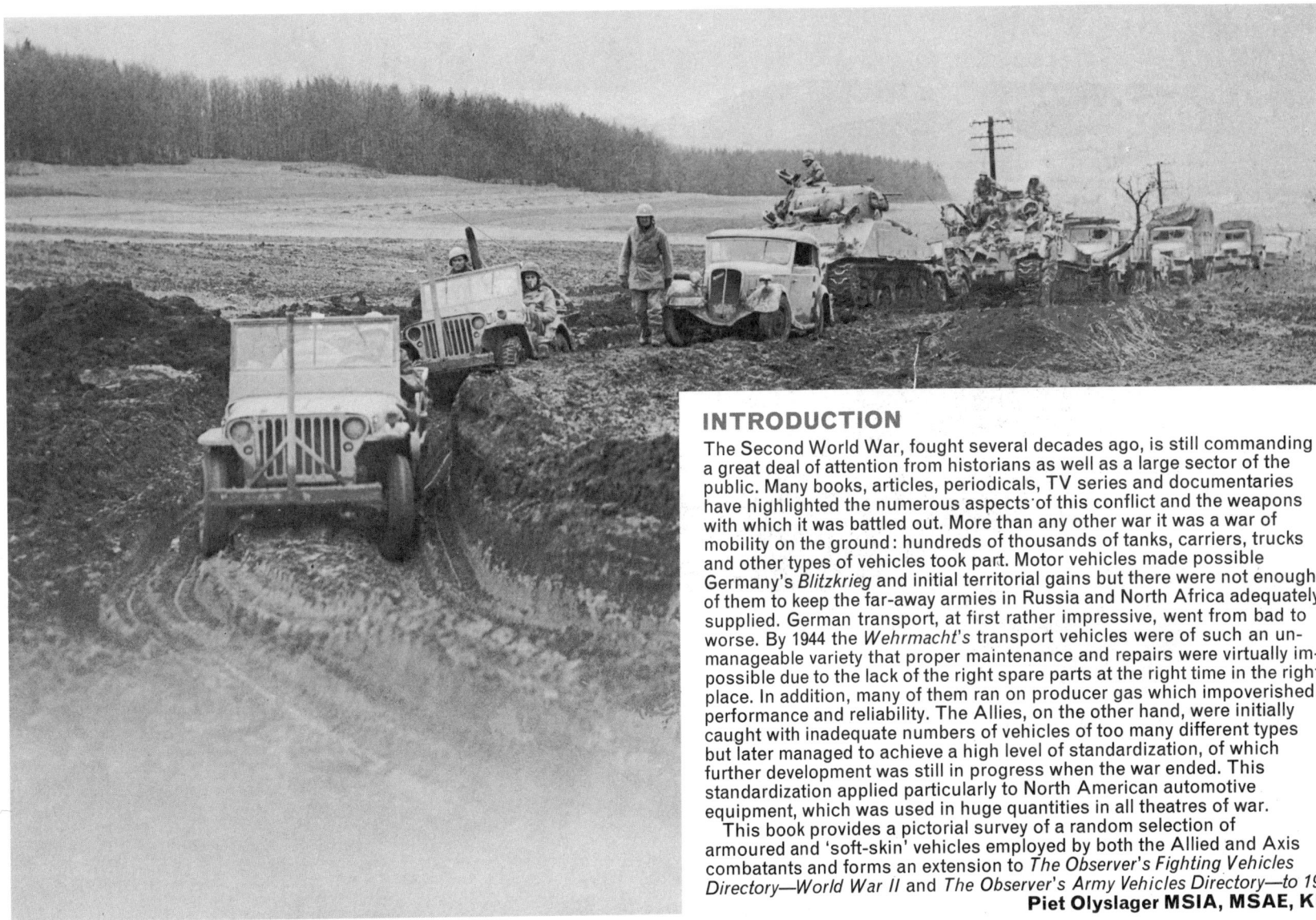

INTRODUCTION

The Second World War, fought several decades ago, is still commanding a great deal of attention from historians as well as a large sector of the public. Many books, articles, periodicals, TV series and documentaries have highlighted the numerous aspects of this conflict and the weapons with which it was battled out. More than any other war it was a war of mobility on the ground: hundreds of thousands of tanks, carriers, trucks and other types of vehicles took part. Motor vehicles made possible Germany's *Blitzkrieg* and initial territorial gains but there were not enough of them to keep the far-away armies in Russia and North Africa adequately supplied. German transport, at first rather impressive, went from bad to worse. By 1944 the *Wehrmacht's* transport vehicles were of such an unmanageable variety that proper maintenance and repairs were virtually impossible due to the lack of the right spare parts at the right time in the right place. In addition, many of them ran on producer gas which impoverished performance and reliability. The Allies, on the other hand, were initially caught with inadequate numbers of vehicles of too many different types but later managed to achieve a high level of standardization, of which further development was still in progress when the war ended. This standardization applied particularly to North American automotive equipment, which was used in huge quantities in all theatres of war.

   This book provides a pictorial survey of a random selection of armoured and 'soft-skin' vehicles employed by both the Allied and Axis combatants and forms an extension to *The Observer's Fighting Vehicles Directory—World War II* and *The Observer's Army Vehicles Directory—to 1940*.

Piet Olyslager MSIA, MSAE, KIVI

The Daimler Scout Car, of which 6,626 were made during 1939–45, was powered by a back-to-front 2½-litre engine, driving through a fluid flywheel with pre-selective five-speed transmission. Final drive was by four individual shafts, one to each wheel. On Marks II (shown) and III the original facility for rear-wheel steering was dispensed with, this having proved a liability in the hands of inexperienced drivers. Similar-looking scout cars were produced in Canada (Ford Lynx). They were widely used for scouting and liaison purposes.

The basic mechanical design can be traced back to 1933 when Nicholas Straussler, a consulting engineer in London, designed a new type of armoured car, featuring rear-engine location, independent wheel suspension front and rear, four-wheel drive and steering and duplicate driving controls at rear. This general configuration was used for several wheeled AFVs (Armoured Fighting Vehicles) produced in subsequent years in various countries (Britain, Canada, France, Germany, Italy, USA).

▽ The German *Wehrmacht* sometimes employed captured British armoured vehicles, such as the two Daimler Scout Cars (Mk IB) shown here in service with the *Afrika Korps*. Lancia in Italy produced a replica, powered by their 8-cyl. Astura engine; a relatively small number of these were made and one has survived, now forming an exhibit in the Italian Military Vehicles Museum in Rome.

▷ In the British Army the Daimler Scout Car saw further active service in Korea. This Mk III, used by an RASC platoon commander, featured a 'Jeep' windscreen and a folding top. The picture was taken near Seoul in 1951.

▷▽ Many Daimler Scout Cars (often referred to as 'Dingos') were auctioned off by the British Government in the 1960s and a fair number were snapped up by military vehicle enthusiasts. This attractively restored specimen was seen in Sussex in 1972.

The Humber Light Reconnaissance Car was front-engined. Shown is the 4 × 2 Mark II.

Humber Scout Car. The Rootes Group in England produced armoured wheeled vehicles of various types, including 4,300 Humber Scout Cars. One is seen here approaching some burning Canadian vehicles on the road to Falaise, Normandy, in 1944.

Humber Scout Car in the south of the Netherlands in December, 1944. Although rear-engined, the Humber chassis, Marks I and II, were of more conventional design than their Daimler counterpart. In the background are a Diamond T recovery unit, carrying a Churchill tank, and a Ford WOA2 Heavy Utility staff car.

Daimler Armoured Car
Both the Daimler Motor Co. and the Rootes
Group produced many thousands of Armoured
Cars, in addition to Scout Cars. Again, the
Daimler (Mark I shown) was the more
sophisticated vehicle, with individual drive
shafts to independently suspended wheels.
Rootes' Humber (see page 26) had 'live' axles
but was more numerous. Although produced by
Karrier Motors Ltd, it was called Humber (both
Rootes Group marques) in order to avoid
confusion with the Bren/Universal Carrier.

The US manufacturers Autocar, Diamond T, White and International Harvester produced a total of over 40,000 half-track vehicles. The first three makes were mechanically similar, being assembled from the same major components. International Harvester, producing mainly for Lend-Lease requirements, used their own engines and other parts. There were many variants, all with superstructure differences, including self-propelled Gun Motor Carriages. Shown is the M2A1, an Armoured Personnel Carrier (APC) for a crew of 10.

Car, Half-Track, M2A1 (Autocar, White).

The US Carrier, Personnel, Half-Track, M3, was a 13-seater APC, produced by Autocar, Diamond T and White. Later models, designated M3A1, featured an armoured machine gun ring mount (or 'pulpit'), like the M2A1 shown opposite. Most types were supplied with either a winch or an obstacle roller at front; the latter is seen here. Large numbers of US half-tracks of various types are still in service with the Israeli armed forces. Most if not all other countries which used them for many years after World War II have replaced them by more modern wheeled or full-tracked types.

Carriage, Motor, 105-mm Howitzer, T19. This was one of the various types of SPs which were based on the standard US half-track chassis. Derived from the M3 APC it carried an M2A1 105-mm howitzer and first saw active service in North Africa. All models produced by Autocar, Diamond T and White had a six-cylinder White Model 160AX side-valve engine of 147 bhp. The T19 SP was produced only by the Diamond T Motor Car Co. of Chicago, Illinois.
(For comprehensive coverage of half-track vehicles the reader is referred to the title HALF-TRACKS in this series.)

The White Scout Car, M3A1, was similar to the US half-track APC in many respects but had four wheels (4 × 4). Photo shows part of the Second Armored Division, US Army, on review at Fort Benning, Georgia. Also displayed are 1940 and 1941 Dodge 4 × 4 command cars and 1941 GMC 6 × 6 cargo trucks. Of the latter, well over half a million were built for the U.S. and allied forces during 1941—45.

US half-track APC in full combat trim after landing on the Normandy beaches in June, 1944. In addition to a British Humber 8-cwt 4 × 4 PU truck (right) this view shows the usual assortment of standard-type US Army vehicles : 'Jeep', Dodge, GMC 'Deuce-and-a-half' and 'Duck', etc. The number on the bumper of the half-track indicates the LST (Landing Ship, Tank) on which it crossed the English Channel.

Germany was the only other country (besides the USA) to produce semi-track vehicles in large quantities during World War II. There were small, light, medium and heavy types and the illustration shows one of the prototypes of the light (1-ton) type, featuring resilient rubber front tyres and towing a 3·7-cm anti-tank gun (*Pak*). Production models had pneumatic front tyres and redesigned bodywork.

▽ Of the *Wehrmacht's* 1- and 3-ton type semi-tracks there was a wide range of armoured editions, in addition to 'soft-skin' tractors and special vehicles. This is a Demag/Wegmann D7p *Sd.Kfz.252* armoured ammunition carrier and command vehicle, utilized by *Sturmgeschütz* units and based on the 1-ton chassis.

▷ A rare variant in the German 3-ton range was this ambulance model, some of which were in service with the German Coast Guard. They were based on the Hanomag *Sd.Kfz.11* chassis.

One of the most common combat vehicles of the German Army was the *Sd.Kfz.251*, armoured variant of the *Sd.Kfz.11*. It appeared in at least 27 versions, varying from the basic *Sd.Kfz.251/1 m.SPW.* APC (shown) to self-propelled artillery mounts.

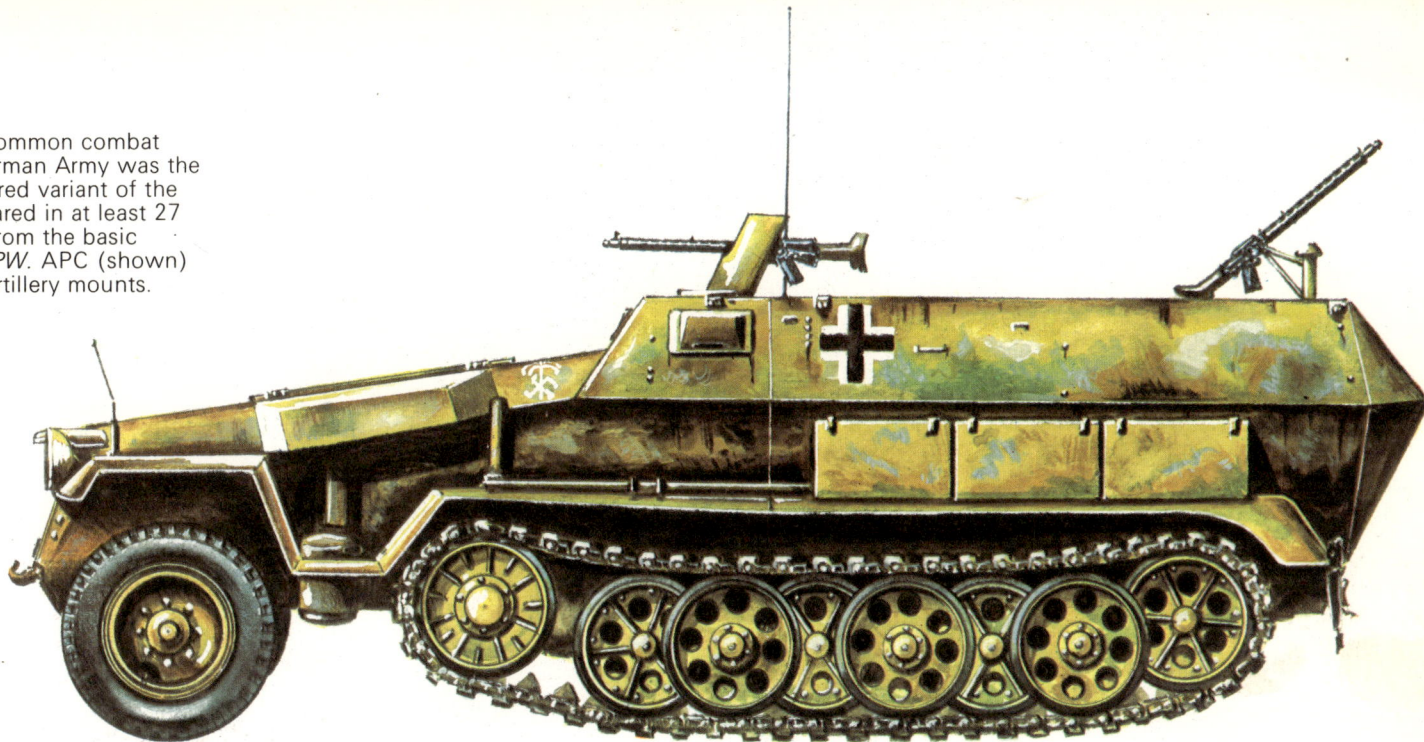

◁ Line-up of *Sd.Kfz.251/1* APCs during manoeuvres. Most types of German semi-tracks were powered by Maybach petrol engines. In the case of the 251 series it was a 100-bhp 4170-cc six-cylinder.

▽ *Sd.Kfz.251/6 mittlerer Kommandopanzerwagen*, a comprehensively equipped command vehicle as used by Rommel (shown), Guderian and other *Panzer* leaders. Note the overhead radio aerial.

Carrier, Bren, No. 2, Mark I. To a large extent the Bren (later Universal) Carrier was to the British what the armoured semi-tracks were to the Americans and Germans. In all cases there were numerous versions and various manufacturers produced them to common specifications. Ford V8 engines of various types were used. Shown is an early type as used at the beginning of the war. For Universal Carrier see pages 15 and 16.

In order to supplement UK production, nearly 34,000 Universal Carriers were produced by Ford of Canada. A small number of these were equipped as 'tank hunters' (Model C21 UCG). The gun was a 2-pounder.

The Ford Motor Co. in the USA also manufactured Universal Carriers (T16) but these were different from the British design in many details. Vehicles shown (T16E2) were used by the Argentine Army for artillery towing. They were also used by the British and, later, by the Swiss Army.

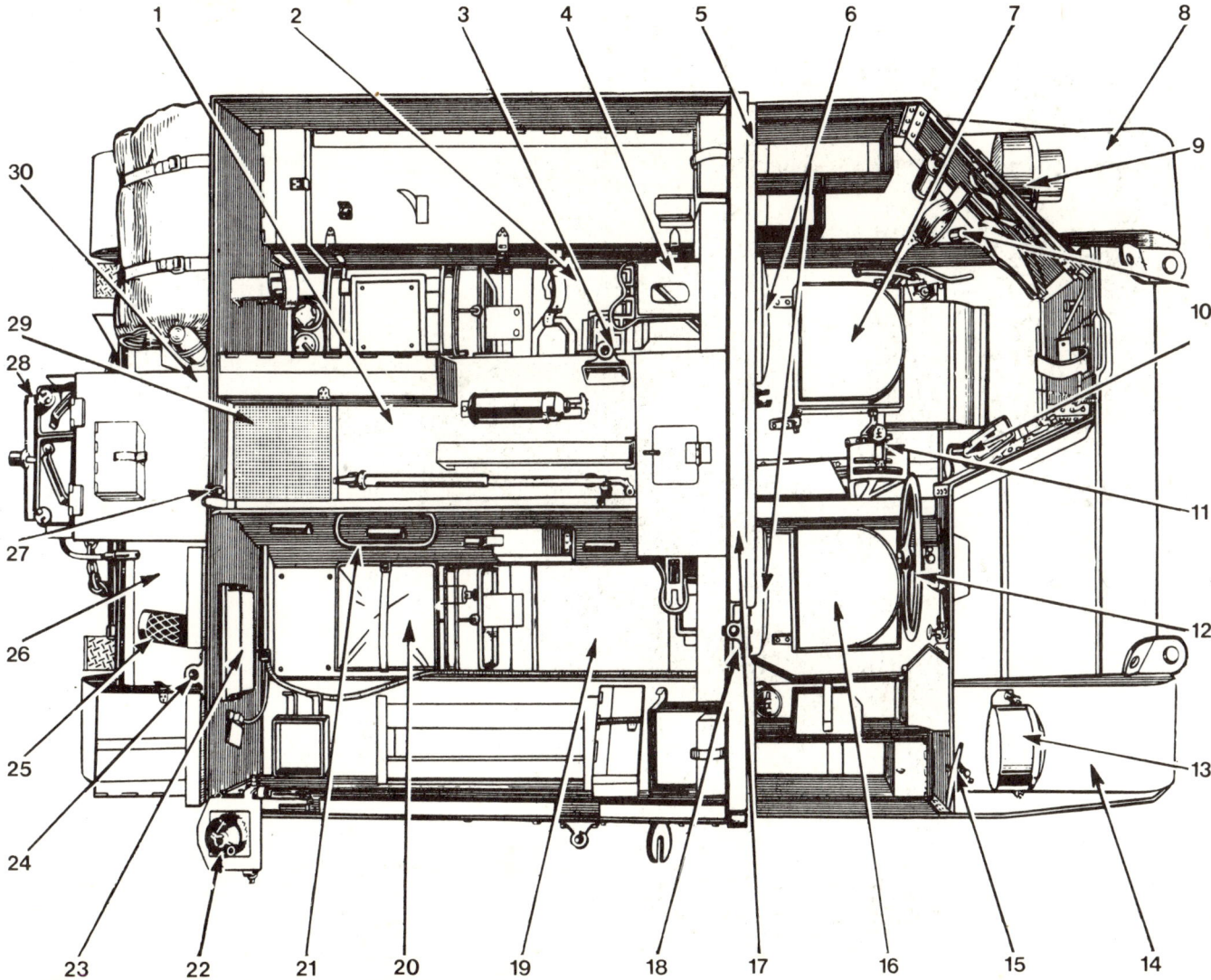

Top view of the Canadian-built Universal Carrier.

Key :

1 Engine cover
2 Floor board
3 Bren gun bracket
4 Battery mounting (under)
5 Division plate
6 Backrests
7 Gunner's seat
8 Mudguard
9 Headlamp
10 Hand grips
11 Gearshift lever
12 Steering wheel
13 Headlamp
14 Mudguard
15 Rearview mirror
16 Driver's seat
17 Firing rest
18 Bren gun bracket
19 Floor board
20 Kneeling/rear seat pad
21 Hand rail
22 Aerial mast mounting
23 Rear seat backrest
24 AA Bren gun mounting
25 Silencer heat screen
26 Axle cover
27 Aerial clamp
28 Petrol/oil containers
29 Radio anti-interference screen
30 Radio protector and mounting

The Universal Carrier (No. 1, Mk I shown) was a light full-track reconnaissance and combat vehicle. Speed and manoeuverability were characteristic. In some vehicles radio equipment provided inter-communication. It was powered by a Ford V8 engine and the power was transmitted to the tracks through a standard clutch, 4F1R transmission and a conventional rear axle with sprockets in place of wheels. Steering was accomplished by a lateral movement of the front bogie assembly for initial steering, and braking of either sprocket to stop the track for acute turns, both of which were accomplished by turning the steering wheel. The hull was bullet-proof plate closely fitted for protection against bullet splash, and riveted at all joints.

The Light Tank, Mark VIC, was the final version of a range of Vickers/Carden-Loyd light tanks produced during the period 1936–40 and powered by a Meadows six-cylinder engine, mounted to the right of the driver. They had Horstmann slow-motion coil-spring suspension, basically similar to that used on the Bren/Universal carriers, but with four road wheels on each side. The Mk VIC had co-axial 15- and 7·92-mm Besa machine guns.

The Light Tank, Mark VII, Tetrarch, was designed by Vickers in 1937 to supersede the Mk VI series. It was not produced in large quantities but in 1943 was adopted for glider-borne operations. The Hamilcar glider (shown) was specially designed and produced to carry the Mk VII and some were landed in Normandy on D-Day. The tank weighed about 16,800 lb and was propelled by a 12-cylinder Meadows engine. Steering was by flexing of the tracks.

In 1941 a Tetrarch was fitted with Straussler swimming equipment, consisting of a collapsible canvas flotation screen and a swivelling propeller/steering device. Successful tests led to the development of Valentine and Sherman DD (Duplex-Drive) tanks.

THE ALECTO SELF-PROPELLED GUN MOUNTING

The Alecto was designed by Vickers to meet the need for a light self-propelled gun for the close support of infantry. It is equally efficient as a gun tractor. Rapid acceleration and great manoeuvrability enable this dual purpose vehicle to maintain a high average speed and all-round performance even under very severe physical and climatic conditions, as recent tests have proved.

ACTING AS GUN TRACTOR · · ·

· · · *UNDER ROUGH CONDITIONS*

Vickers-Armstrongs Limited

The Mk VII Tetrarch light tank was followed by the rather similar Mk VIII Harry Hopkins, which was also designed by Vickers-Armstrongs Ltd. The Harry Hopkins had a different hull and turret. It was not used in action. This advertisement shows the Alecto SP gun mounting, which was derived from it. A dozer version also appeared.

The Light Tank, M3, was produced in the USA from 1941 until 1943 when it was supplemented by the M5 series which was basically similar. Of both there were several variants and derivatives. The M3 shown was powered by a radial 7-cylinder Continental petrol engine (a small number were fitted with a radial 9-cylinder Guiberson diesel). The British used the M3 first in North Africa, later also elsewhere, and named it the (General) Stuart but it was popularly known as the 'Honey'.

Light Tank, M5, as owned by a British military vehicles collector. It was originally used by the British Army (as Stuart VI) who later removed the turret; such vehicles were employed for reconnaissance, artillery towing and other roles. The turret shown is a replica. The M5 was propelled by twin Cadillac V8 engines, driving through Hydramatic automatic transmissions.

The Light Tank, M24, known as the Chaffee, succeeded the American M3 and M5 series and made its appearance in 1944. Like the M5 it had twin Cadillac V8 engines with automatic transmission. There were, again, several derivatives, including Gun and Howitzer Motor Carriages.

The Chaffee tank was employed by many nations until long after the war. Specimens shown belonged to the Japanese Self Defence Forces.

This advertisement features the American M18 'Hellcat', which, although looking like a tank, was in fact a high-speed Gun Motor Carriage (SP Gun). It was successfully employed as a hit-and-run tank destroyer in Italy and NW Europe during 1944–45.

Introduced in 1934 the German *Panzerkampfwagen I, Ausf. A* light tank was widely used in the early stages of World War II, although it was intended mainly for training. The engine was a Krupp 57-bhp 4-cyl. 'boxer' with a five-speed gearbox. *Ausführung B* (second version) had a six-cylinder Maybach engine, a longer hull, five road wheels and other improvements. The aircraft are *He 51* fighter biplanes.

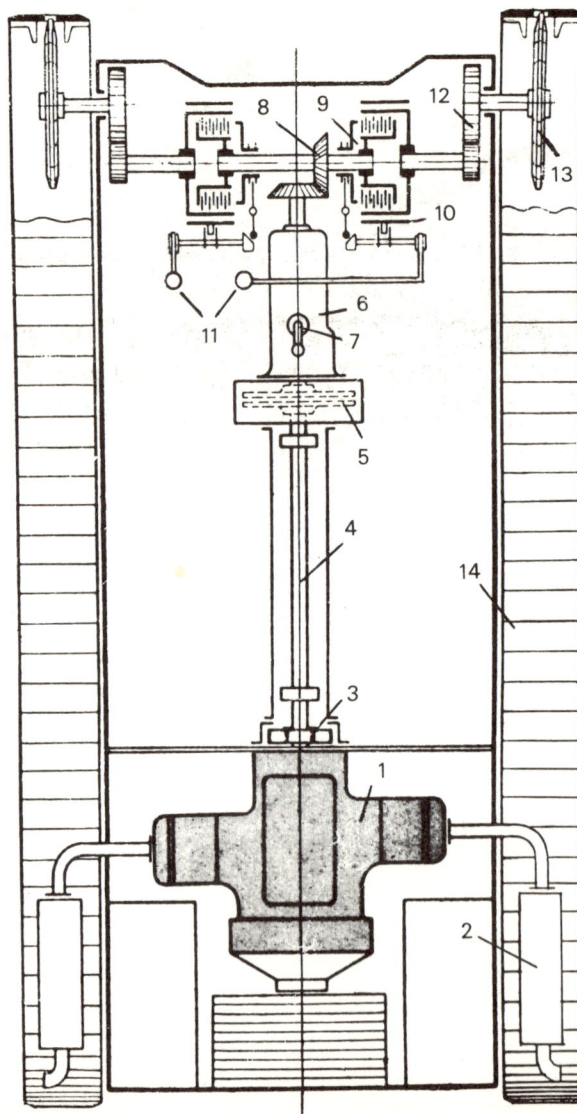

Key:
1 Engine (Krupp M304, petrol)
2 Exhaust silencer
3 Gear reduction unit
4 Propeller shaft
5 Main clutch
6 Gearbox
7 Gearshift lever
8 Bevel gears
9 Steering clutches
10 Clutch brakes
11 Steering levers
12 Final drive
13 Drive sprocket
14 Track

Power train of the *Pz.Kpfw.I* (*Sd.Kfz.101*).

▷ Front view of the *Pz.Kpfw.I, Ausf.A*. The coil springs of the leading road wheels are just visible ; the other wheels were leaf-sprung. These light tanks were armed with two 7·92-mm machine guns and weighed 5200 kg (*Ausf.B.* weighed 5600 kg).

Principal German tanks at the beginning of the war :

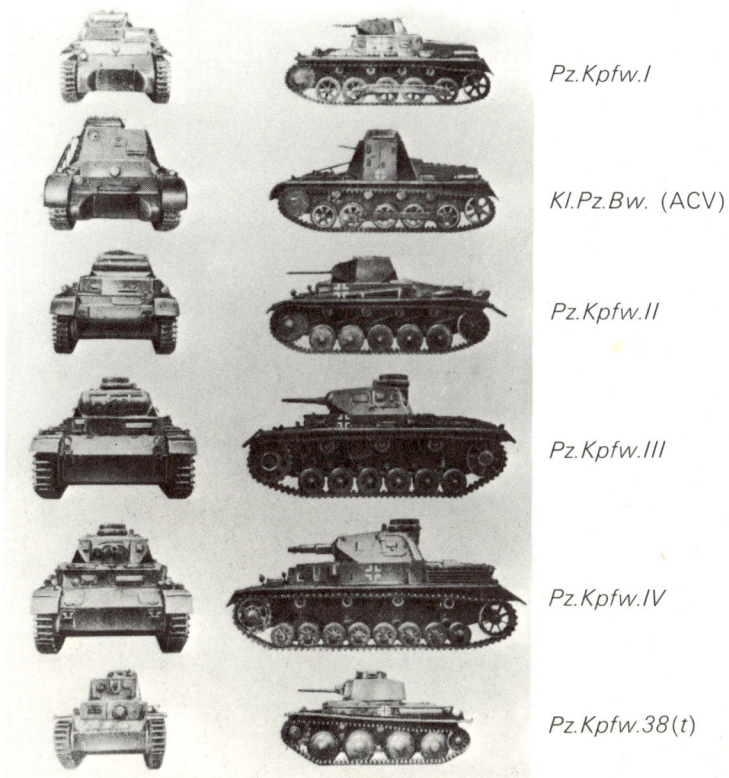

Pz.Kpfw.I

Kl.Pz.Bw. (ACV)

Pz.Kpfw.II

Pz.Kpfw.III

Pz.Kpfw.IV

Pz.Kpfw.38(t)

▷ *Pz.Kpfw.I, Ausf.B* (right), accompanied by *Pz.Kpfw.II* (left) and *Pz.Kpfw.IV* (background) during manoeuvres in Germany in the late 1930s.

Panzerkampfwagen II, Ausf.B (Sd.Kfz.121) was developed in the late 1930s and widely used during the early years of the war. There were several variants, including SP guns. The rear-mounted power unit was a 140-bhp water-cooled Maybach petrol six-cylinder, driving through a 6F1R transmission. The road wheels, five on either side, were independently suspended and leaf-sprung.

The Matilda IV CS was basically the standard Matilda Infantry Tank but the 2-pounder gun was replaced by a 3-inch howitzer. The Matilda was one of the best known tanks during the early war years and was employed in North Africa, Eritrea, Crete and Malta. They were, in addition, supplied to Australia and the Soviet Union. Early models had twin AEC diesel engines of 87 bhp each, in later Marks these were replaced by two 95-bhp Leyland diesels. In North Africa their supremacy ended when the German 88-mm gun made its appearance.

Armoured vehicles awaiting issue to Armoured Division at a Middle East Base Depot during the Desert War. On the left is a Cruiser Tank, Mk VIA, Crusader, behind it a Cruiser Tank, Mk IVA, and in the background a row of Valentine Infantry Tanks. Note the supplementary fuel tanks. On the right and far left are Humber Armoured Cars, armed with 15-mm and 7·92-mm Besa machine guns.

Over 600 Valentine tanks were equipped with Straussler DD swimming equipment. Shown under construction is a version for steep ramp launching. The folding canvas screen, not yet installed here, was erected by inflating the two sets of air tubes. The propeller was driven from the tank's main engine.

Valentine DD tank leaving the ramp of an LCT (Landing Craft, Tank) and entering the sea about 3000 yards from shore, during exercises in the UK.

Prior to reaching the shore the flotation screen was quickly lowered by deflating the air tubes. A few Valentine DDs were used operationally in Italy. Most DD tanks were Shermans, converted in Britain and the United States. On the Sherman DD the propellers were driven from the rear ends of the tracks. Steering in water was accomplished by swivelling the propellers.

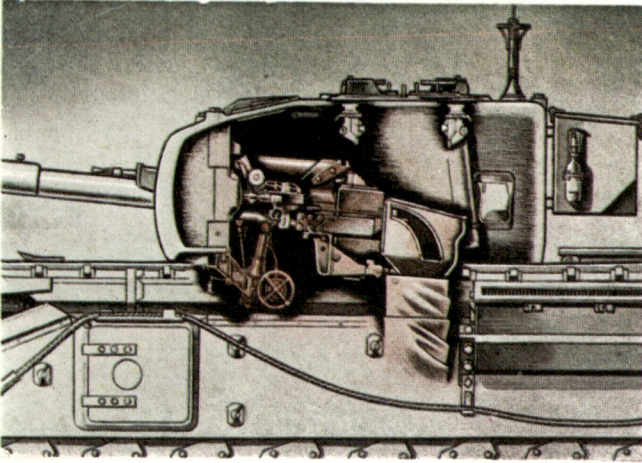

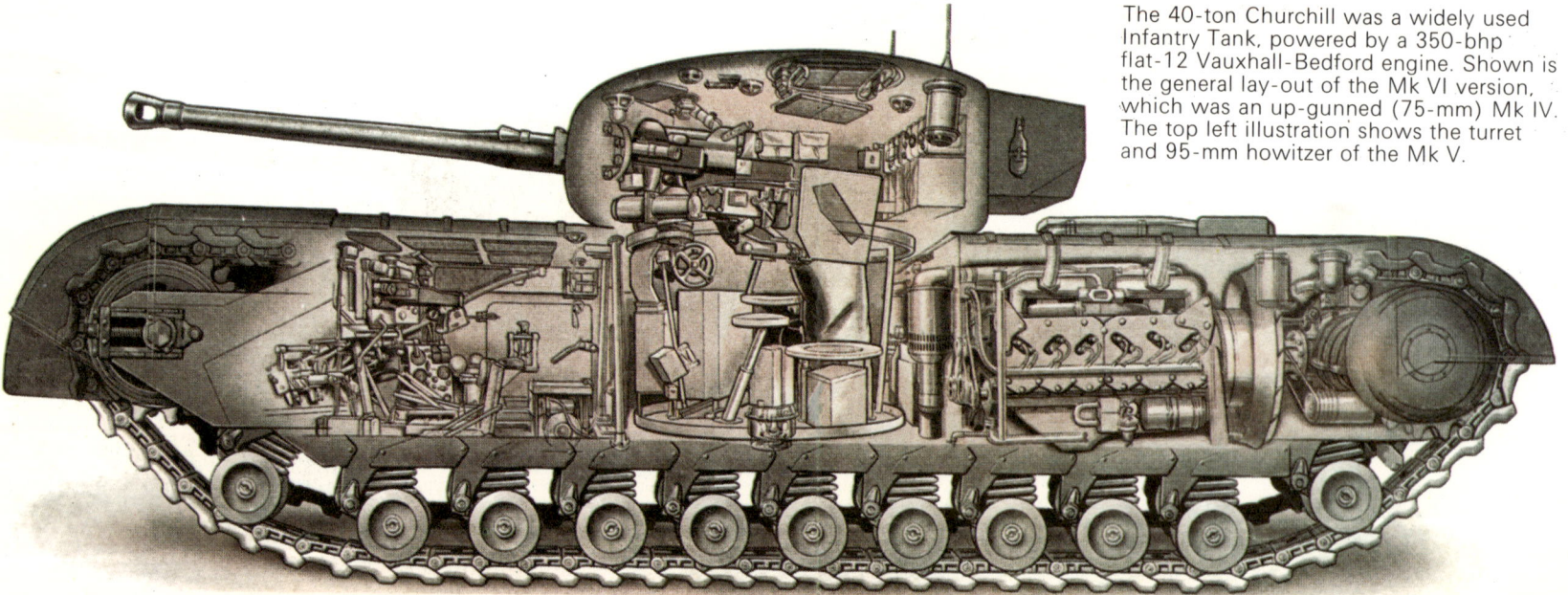

The 40-ton Churchill was a widely used Infantry Tank, powered by a 350-bhp flat-12 Vauxhall-Bedford engine. Shown is the general lay-out of the Mk VI version, which was an up-gunned (75-mm) Mk IV. The top left illustration shows the turret and 95-mm howitzer of the Mk V.

The 90-mm Gun Motor Carriage, M36, was a tank destroyer and one of numerous AFVs based on the famous American Sherman medium tank chassis. Although having the appearance of a tank, this was, in fact, a self-propelled gun (SP Gun). The hull was similar to that of the 3-in Gun Motor Carriage, M10A1. The M36B1 had the same gun as the M36 but the hull of the Medium Tank, M4A3. All three had a Ford V8 Model GAA 1100-cu. in petrol engine of 450 bhp, driving through a five-speed transmission.

Churchill VII
Infantry Tank, a much-improved version compared with the earlier Marks. It had a redesigned hull and turret (with 75-mm gun) and other modifications. Altogether over 5,600 Churchill AFVs were built by a group of nine manufacturers, all under Vauxhall Motors' 'parentage'.

Sixth South African Armoured Division troops with predominantly US-built vehicles congesting a main road into Bologna, Italy, in April, 1945. Shown are M4 series Sherman medium tanks (left and background), M10 Gun Motor Carriage (centre foreground), 'Jeeps', Humber and Daimler Scout Cars, Universal Carrier, etc. Note the many items of additional equipment carried on the vehicles.

△ (General) Sherman M4A1 medium tanks made their battle debut when operated by the British Eighth Army at El Alamein in the Western Desert in 1942. In tank battles during the Yom Kippur War, not very far from here but more than 30 years later, Sherman tanks again took part, albeit in modified form.

△ ▷ Sherman tanks were powered by various types of engines, including this 30-cyl. Chrysler multi-bank radial unit which combined five Dodge 6-cyl. engines. It produced 370 bhp and was used in the M4A4 version. Other power units used in the M4 series were: 350-bhp Continental 9-cyl. radial (M4, M4A1), 375-bhp GM twin 6-cyl. diesel (M4A2), 450–500-bhp Ford V8-cyl. (M4A3).

American, British and Canadian vehicles, a common sight on the Normandy front in 1944. Photo shows British and Canadian troops in Putanges, south of Caen. The vehicles are a US-built Sherman with 17-pdr gun (British modification, known as 'Firefly'), a British Ford WOA2 Heavy Utility staff car and a Canadian Chevrolet 15-cwt 4 × 4 GS truck. Just visible on the left is the front end of a British Standard 12 HP Light Utility.

The Sherman Medium Tank, M4A2, differed from the basic M4 in that it had twin GM Series 71 six-cylinder two-stroke diesel engines instead of the M4's air-cooled radial Continental Model R-975. This cut-away drawing shows an early production M4A2 with 75-mm gun; late production vehicles had a more effective 76-mm gun. The power unit drove the forward-mounted gearbox-cum-final drive via a universally jointed propeller shaft.

All models had controlled differential steering and a crew of five. Shown is the first-type bogie, with overhead track return rollers. Second-type bogies had these rollers offset to the rear and a track skid on top of the bracket. Much later these bogies, which had vertical volute springs, were replaced in production by a new horizontal volute spring suspension (HVSS). From 1942 the American industry turned out over 40,000 of the ubiquitous Shermans and their derivatives.

The Sherman M4A1E9 featured experimental spaced-out suspension and extra-wide tracks, i.e. extended end connectors or 'grousers', developed by Chrysler's Detroit Arsenal. It was one of many experimental versions developed or modified during the war.

◁▽ A Sherman that was not a tank. This inflatable 'tank' was produced in some quantity by Dunlop and was one of a range of models used, reputedly, during the break-through at the Falaise Gap and on the Rhine Crossing. Made from rubberised cotton yarn these devices replaced earlier wooden and canvas decoys.

▽ A surviving Sherman standing guard over the D-Day landing beaches at Arromanches, Normandy.

◁ After the war many countries equipped their armed forces with Shermans. This bridge-laying variant was used by the Japanese Self-Defence Forces.

△ During 1942–3 Argentina designed and produced 16 'Nahuel' (Jaguar) DL43 35-ton medium tanks, which were clearly patterned on the Sherman. Also visible in this photo, which shows a parade in mid-1944, is a Belgian-made FN armoured motorcycle combination.

◁ In 1942 Australia produced 66 Cruiser Tanks, ACI (later named Sentinel), which were not unlike the US Medium M3/M4 type tanks. They were powered by triple Cadillac V8 engines, totalling 24 cylinders! Canada also produced similar Cruiser Tanks, namely the Ram and the Grizzly; the latter was the Canadian version of the Sherman M4A1.

Soviet KVI heavy tank, introduced in 1939/40. Powered by a V12 diesel engine of 550 bhp it was armed with a 76·2-mm gun and two MGs. Frontal armour was up to 100 mm thick. The tank weighed 42,000 kg and was about 6·70 metres long. In this typical Nazi artist's impression, presumably based on an actual encounter, a KVI suddenly appears from the right and pushes the leading *Panzer* sideways into a house. In rather emotional words the original German caption then narrates how, during the few moments this takes, the second *Panzer* machine-guns the Soviet's vision slits, causing the latter to lose direction and bury itself in another building. "The *Blitzkrieg* continues . . . ."

Soviet medium T34, one of the best and most successful tanks of the war. It remained in use for many years, albeit with various modifications. The original 76·2-mm gun (shown) was later replaced by an 85-mm in a redesigned turret (T34/85). Like most of the Red Army's tanks it was powered by a V12-cylinder diesel engine.

During the latter part of the war the Soviets' standard heavy tank was the Stalin (JS, or IS). Illustrated is the second version, which was generally similar to the original. It had a long 122-mm gun and three MGs. The third and final edition (JS or IS III) had a redesigned hull front and turret. They were also used as the basis for various types of self-propelled guns.

The German *Pz.Kpfw.VI 'Tiger'* first saw active service in the Leningrad area in 1942 and was equipped with the fearsome 88-mm gun. They were also used, briefly, in North Africa and subsequently in Italy and NW Europe. About 1,350 were built, including some *'Sturmtiger'* self-propelled 38-cm mortars. This cut-away view shows a *'Tiger I'*, *Ausf.E*. The rear-mounted 700-bhp Maybach V12-cyl. engine drove the front sprockets via an Olvar pre-selector transmission. The tank weighed about 55,000 kg and on roads could attain 38 km/h. Frontal armour was 100 mm, side armour 80 mm (top) and 60 mm (bottom). Overall length was 6·20 m (incl. gun 8·24 m).

The improved *'Tiger II'* or *'Königstiger'* appeared in 1944. The Allies called it the 'King Tiger' or 'Royal Tiger'. At 69,700 kg it was the heaviest operational tank made by any of the combatants in WWII.

Pz.Kpfw.III, Ausf.L (Sd.Kfz.141) leaving the works. This version featured 20-mm spaced armour across the front of superstructure and turret. The 'Panzer III' was powered by a 300-bhp Maybach V12-cyl. engine, driving through a Variorex pre-selector transmission with 10 forward and 4 reverse speeds.

German tank workshop, showing, in the foreground, one of the rare heavy Nb.Fz. tanks. This type, a few of which were used in Norway in 1940, had auxiliary machine-gun turrets at front and rear. In the background are Sd.Kfz.250, 251 and other AFVs.

'Tiger I' surviving at Aberdeen Proving Ground, Maryland, USA. It was captured and shipped over from Tunisia. Note the large 'interleaved' road wheels. Behind it rests a Pz.Kpfw.IV ('Panzer IV'), captured in Normandy, France. In the background are an Italian 75-mm SP howitzer (on M13/40 tank chassis) and a Japanese CHI-HA medium tank, all of WWII.

The *Wehrmacht* used considerable quantities of AFVs made in occupied countries, particularly Czechoslovakia. This is a reworked Czech LT38 with Russian-made SP 76·2-mm AT gun (*Pak*).

The Italian Army in World War II used mainly light (L series—*leggero*) and medium (M series—*medio*) tanks. Heavy types (*pesante*) did exist but were not used in quantity. Shown is a *Carro Armato* L6/40 as produced by Fiat and Ansaldo. The chassis was also used for SP Guns (*Semovente*), and was easily distinguishable by its two bogie units, sprung on prominent torsion-bar suspension arms. Armament consisted of a 20-mm gun and one MG. The engine, at rear, was a Fiat/Spa petrol four-cylinder of 68 bhp. Note jack carried below right headlamp.

Semovente Commando M42, a command vehicle based on the chassis of the M14/41 (which was the late production version of the M13/40 medium tank, *see* following page). This chassis had eight dual road wheels suspended in pairs on leaf-sprung cranks, and a Fiat/Spa V8 diesel engine of 145 bhp.

Italian *Carro Armato Medio M13/40*. This tank formed the backbone of the Italian armoured forces during 1940—43. Ansaldo-Fossati of Genova-Sestri produced about 2000 units of this and the improved M14/41 version. There were also self-propelled artillery and other variants. The M13/40 carried a 47/32 gun, three 8-mm MGs and a crew of four.

Japanese Medium Tank, Type 97, *CHI-HA*. Designed in 1937 this was one of the best tanks used by the Japanese armed forces. There were several variants, including a command version. Hull and turret were built up from riveted armour plates and the original 57-mm gun was later superseded by a 47-mm QF type. The engine was a V12 diesel of 170 bhp.

Berliet *Porte-Char*, a 12-ton tank transporter, shown loading an infantry medium tank of the *Char D1* type. The loading hoist, which lifted the tank from the ground on to the truck, was actuated by screw spindles in the load platform sides; the spindles were worm-driven from a PTO (power take-off) on the truck's transmission. The tandem-drive truck had a GVW (gross vehicle weight) rating of 20,000 kg; the engine was a six-cylinder diesel. Additional 6 × 4 tank carriers (but not of the self-loading type) were ordered from the US in 1939 but were diverted to Britain when the war broke out. Much of France's military equipment was destroyed or fell into enemy hands.

Harley Davidson motorcycles were widely used by the Allied forces. This is a Model WLC of the Royal Canadian Army Service Corps.

German BMW solo machine, one of many types of this make used by the *Wehrmacht*.

Typical application of Italian Moto Guzzi 500-cc motorcycle with 'sprung frame'.

◁ During the 1930s the German *Reichswehr* and *Wehrmacht* used many *Kübelwagen* (bucket seat cars) on commercial car chassis of various makes. Shown are a Mercedes-Benz Stuttgart (front) and a Wanderer W11.

◁ Wanderer *Kübelwagen* under test. It was officially known as *mittlerer geländegängiger Personenkraftwagen* (medium cross-country passenger car). There were also light and heavy types.

△ From about 1937 the improvised 4 × 2 *Kübelwagen* types were superseded by light, medium and heavy standard 4 × 4 models of highly sophisticated design. Shown is the medium type (*m.E.Pkw.*), produced by Auto-Union/Horch and Opel with V8 and 6-cyl. engine respectively.

▷ Heavy standard car (*s.E.Pkw.*), made by Auto-Union/Horch and Ford-Köln, both with V8 engine.

The 'Jeep' in action with the British Army at El Alamein in 1942.

The famous American 'Jeep' was employed in vast numbers by all the Allied forces for a great many purposes. Of nearly 640,000 produced (by Willys-Overland and Ford) it is estimated that about a quarter of a million were lost in combat. The remainder saw many years of further service, both with the armed forces of numerous countries and in Civvy Street. (For its full story the reader is referred to THE JEEP in this series.)

Beautifully preserved 'Jeep', owned by Mr Fernand Tiquet of Belgium

Hay snot face

△ The *Wehrmacht* made wide use of motor-cycles with sidecars, exemplified by this 500-cc twin-cylinder Zündapp. Later models (BMW and Zündapp military designs) had a 750-cc engine and power transmission to both rear and sidecar wheel.

△ ▷ One of the most common German vehicles in World War II was the Volkswagen/KdF *Typ 82 Kübelwagen*. It was, like the US 'Jeep', a jack-of-all-trades, although in design the two were vastly different.

▷ In addition to the *Kübel*, the KdF (Volks-wagen) works produced the amphibious *Typ 166 Schwimmwagen*. Unlike the *Kübel* it had four-wheel drive and, particularly on land, it had a very good performance.

The well-known Austrian Steyr concern, old-established armament and motor vehicle producers, supplied various types of wheeled and tracked vehicles to the *Wehrmacht*. Most numerous was the Model 1500A

chassis, which appeared mainly with passenger and truck bodywork. At one time it was co-produced by Auto-Union in Germany. Shown are a personnel carrier and a fire-fighting derivative.

The bulk of the *Wehrmacht*'s transport vehicles were civilian types of various origins. This typical assortment was part of a large number seized by the US First Army near Antweiler, Germany. Left to right: Peugeot, Mercedes-Benz, Peugeot, Citroën, Mercedes-Benz, Renault, Tatra, Citroën, etc.

The US Army's standardized military field cars were the Willys/Ford 'Jeep' and the Dodge 'Beep'. Examples of both are seen here near Zell-am-See, Austria, in May, 1945. Note the (German) truck which missed the hairpin bend and landed on its side.

Japan's military field car was the 'Black Medal' Scout Car, Type 95, a design of 1935 made by Kurogane. An air-cooled 1·4-litre V-twin motorcycle-type engine drove all four wheels. Some were later used by the French forces in Indo-China.

Toyota, in 1943–44, produced a number of 4 × 4 truck-based amphibians, designated SUKI, for the Japanese forces. They weighed about 4000 kg and could carry two tons. The company, for the war effort, produced mainly conventional cars and trucks.

The US automotive industry produced amphibious trucks on the 'Jeep' 4 × 4 and GMC 'Deuce-and-a-half' 6 × 6 chassis. The latter, known as the 'Duck' (GMC Model DUKW-353), was by far the most numerous and in some countries is still in service. After the war the Soviet Union copied both the 4 × 4 and the 6 × 6 versions with only detail differences.

The 'Duck' in use by the German *Bundeswehr* in the 1960s. Numerous countries employed these very useful vehicles, which were intended mainly for ship-to-shore ferrying of troops and supplies. The tyre pressures could be regulated by the driver to suit the prevailing sand/road conditions.

◁ 'Ducks' of the British Army, with their canvas tilts erected, and their smaller cousin, the 'Seep' (seagoing 'Jeep'; one of the nicknames of the amphibious ¼-ton 4 × 4). (IWM photo H38993.)

◁ ▽ Experimental 'Duck'-based aerial ladder, produced by Merryweather, fire-fighting equipment manufacturers in Great Britain. It was designed for use in the Normandy landings, enabling troops to climb high cliffs.

▽ The conventional 'Land Jeep' had a fording ability of up to 25 inches but, suitably equipped, could cross waterways completely submerged. In the latter case snorkels ensured the engine's 'breathing'. Note the overhead guide tape, which indicated the shallowest wading course.

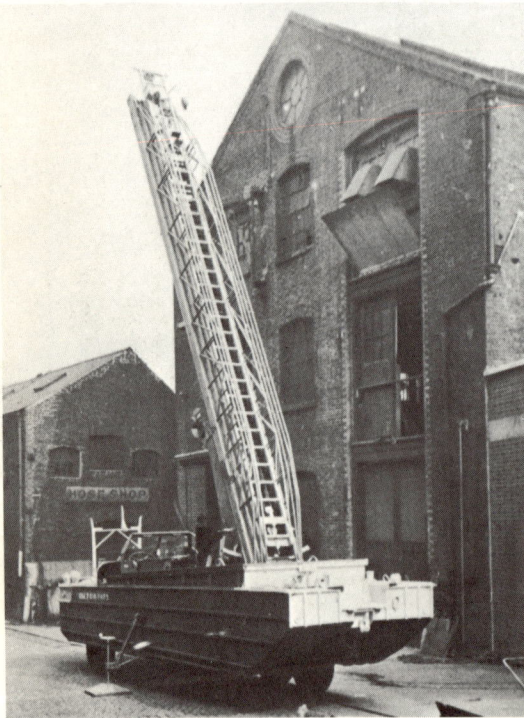

Dodge field ambulances and other transport of the US Fifth Army after crossing a pontoon bridge over the River Po, Northern Italy, in April 1945.

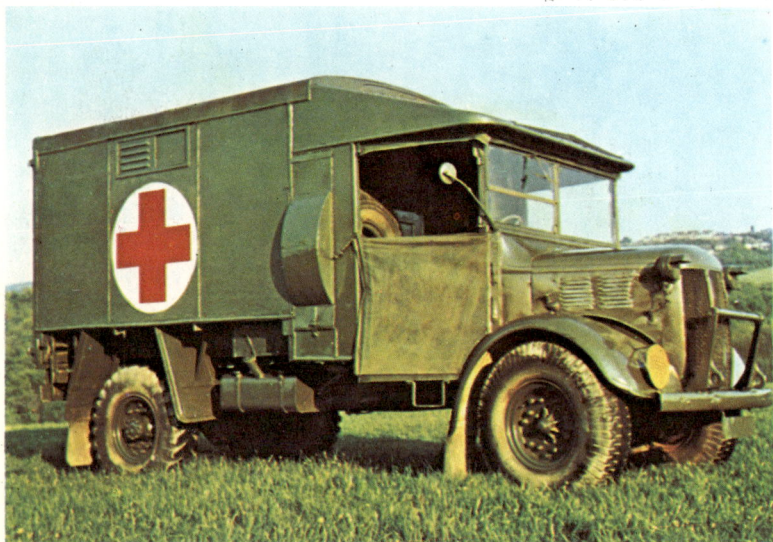

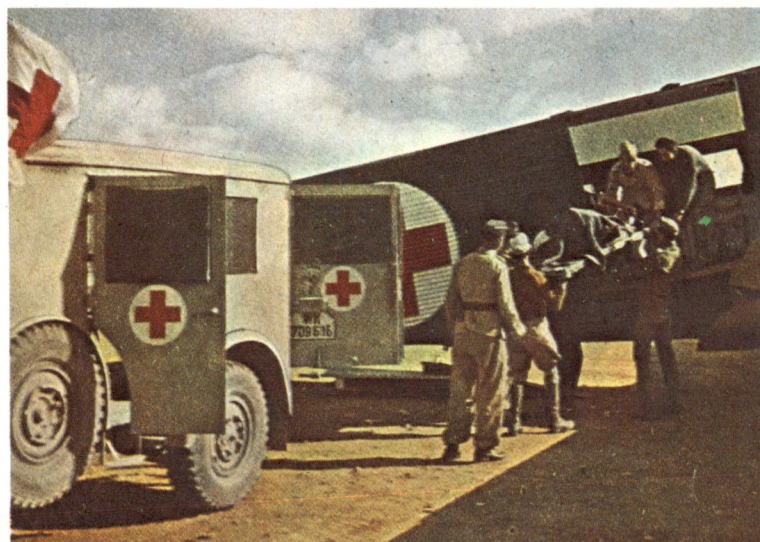

◁△ Best known of British Ambulances in World War II was the ubiquitous Austin K2 which carried four stretchers. It was also supplied to other nations, initially under lend-lease arrangements, even to the US Army. The *Wehrmacht* used a fair number of captured units.

◁ Standard Canadian-built field ambulance was the Chevrolet C8A HUA, a variant of the 8-cwt 4 × 4 Heavy Utility model in the range of Canadian Military Pattern vehicles. This specimen was restored by the Editor in the late 'sixties from a dilapidated breakdown truck.

△ The German *Wehrmacht* employed a large variety of ambulances. Shown is a standard type *Kfz.31*, which was based on the standard heavy car (*s.E.Pkw.*) chassis (see also page 43), in North Africa.

△ Britain's only Field Ambulance was based on the standard Humber four-wheel drive chassis. The bodywork was made by Thrupp & Maberly.

△ ▷ RAF Mountain Rescue Unit, using a Humber and a Willys 'Jeep' The 'Jeep' was, in fact, frequently used as a front line ambulance in most theatres of war, often with improvised stretcher equipment.

▷ The Humber 8-cwt 4 × 4 chassis formed the basis for various vehicle types, varying from staff cars to armoured reconnaissance vehicles. This Radio Recording Van, however, was a straight conversion of the ambulance and was one of several used by the BBC in battle zones in Italy and elsewhere.

At the beginning of the war the British Army used a variety of American Mack and White tank carriers. These were basically heavy commercial type 6 × 4 trucks. Shown is a White Model 920 18-tonner, loaded with a Cruiser Tank (a reworked Mk IIA). (IWM photo MH8327.)

Scammell 30-tonner, sometimes called 'Prairy Schooner', carrying a Cruiser Tank Mk IVA minus its armament (2-pdr gun and co-axial MG).

Standard British-made tank transporter/ recovery vehicle was the 6 × 4 Scammell Pioneer TRMU/30 with Scammell TRCU/30 semi-trailer. This combination was rated for 30 tons and had a Gardner 6LW diesel engine of 102 bhp. A power-winch for 8-ton line pull was mounted on the tractor unit.

For additional heavy tank transporters the British Army turned to the United States where the Diamond T/Rogers 45-ton M19 tractor-trailer combination was designed to meet their requirements. Shown is the early Model 980 tractor. It had a 201-bhp (gross) Hercules DFXE diesel engine and 12 forward speeds. (IWM photo STT 6878.)

Diamond T Model 981 Recovery Tractor, which usually had a soft-top cab, with Rogers trailer and Cromwell Cruiser Tank about to be loaded.

The Diamond T in British service also towed other types of trailers. Illustrated is a British Mk I 40-tonner, produced by Crane and Dyson, carrying a Churchill tank. Four- and five-axle trailers were also made, for larger payloads.

Standard British Heavy Breakdown Tractor was the Scammell SV/2S, which had a light sliding-jib crane but tremendous pulling and winching power. It was basically similar to the tank transporter tractor (see page 54) and the Model R100 heavy artillery tractor. All were of the well-known Pioneer type, with exceptional front and rear wheel articulation. They were popularly known as 'Coffeepot Scammells', on account of the raised central water pot on the top of the radiator.

The Scammell's rear bogie articulation. This particular vehicle was experimentally fitted with a driven front axle, providing six-wheel drive. (For further coverage of Scammells see the title SCAMMELL VEHICLES in this series.)

Diamond T/Holmes Wrecker of the US Army (see also following page) with Dodge 1½-ton 4 × 4 truck on suspended tow.

Probably the best-known and most-liked recovery vehicle to emerge from World War II was the Diamond T 969 series 4-ton 6 × 6 with twin-boom Holmes wrecker equipment. It first appeared in 1941 and from 1943 was also supplied with soft-top cab (shown). It was powered by a six-cylinder Hercules petrol engine. One is seen here putting an overturned GMC 2½-ton 6 × 6 cargo truck back on its wheels somewhere on the 'Red Ball' express route. 'Red Ball' convoys carried supplies from ports and beachheads to troops on the NW European front lines following the D-Day invasion. The routes were restricted to this purpose and the convoys covered up to 900 miles daily.

Along the 'Red Ball' express route were truck company repair and maintenance depots whose sole task was to keep the convoys' wheels rolling.

The *Wehrmacht* used heavy trucks and four-wheeled low-loader trailers (*Sd.Ah.115*) to carry light tanks and large 20- and 23-ton eight-wheeled trailers for heavier types. Illustrated is the 23-ton *Sd.Ah.116*, loaded with a *Pz. Kpfw.IV* and towed by a Famo *Sd.Kfz.9* 18-ton *Zugkraftwagen*.

NOTE: *Sd.Ah.* stood for *Sonderanhänger* (special trailer), *Sd.Kfz.* for *Sonderkraftfahrzeug* (special motor vehicle). Other common German abbreviations included *Pkw.* (*Personenkraftwagen*—passenger car) and *Lkw.* (*Lastkraftwagen*—truck), both preceded by *E.* (*Einheits*) for standardized types. *Pz.Kpfw.*, *Pz.* and *Panzer* were short for *Panzerkampfwagen* (tank).

The German eight-wheeled tank-transporter trailers had a detachable rear bogie on which there was a soft-top cab for the rear steersman. Picture shows *Sd.Ah. 116* with rear bogie removed, ready for tank to be loaded.

Wreckers were not numerous in the *Wehrmacht*. Captured or confiscated civilian models outnumbered military types. This is an ex-Soviet Army GAZ-AAA on the Russian front in October 1941. The GAZ-AA (4 × 2) and AAA (6 × 4) 'Russki Fords' were patterned on the US Ford AA truck.

◁ An artillery tractor widely used by the Italian Army was the *Trattore Leggero TL37*, made by Spa. There were several body types on this all-wheel drive chassis which was made with solid and pneumatic tyres.

◁ ▽ The *Wehrmacht*, at the beginning of the war, used horses as well as very sophisticated semi-track tractors for artillery towing. The tractors shown are of the *Sd.Kfz.8* 12-ton type.

▽ Designed by Porsche, produced by Skoda and intended for the Eastern front, the RSO tractor (*Radschlepper Ost*) saw limited service on the Western front. This one was found abandoned in Normandy in August 1944.

◁ Most numerous of the *Wehrmacht*'s trucks was the Opel Blitz 3-ton 4 × 2 (*S-Typ*); some 70,000 of these were in use. In addition Opel built about 25,000 similar trucks with four-wheel drive (*A-Typ*). Many body types were fitted. Picture shows *S-Typ* cargo truck being unloaded in Norway.

◁ ▽ Opel Blitz *S-Typ* trucks of the Afrika Korps during a roadside stop.

▷ Many thousands of confiscated civilian trucks were also used by the *Wehrmacht*. Chevrolet (shown, 1940) and Ford were particularly popular but as the war drew on virtually anything on wheels was impressed into German service.

▷ ▽ During the war several German truck manufacturers produced conventional medium trucks for the war effort. These are Klöckner-Deutz/Magirus *A-Typ* (4 × 4) three-tonners, which also existed in *S-Typ* (4 × 2) form.

Typical display of German *Wehrmacht* vehicles in the late 1930s. The picture was taken in Berlin and shows mainly Krupp *Protzkraftwagen* *Kfz.69* prime movers with AT guns (*Pak*). In the foreground are Stoewer *Kfz.15* medium cars and Horch *Kfz.17* signals vans.

Abandoned Nazi equipment in Paris, after the capital's liberation on 25 August, 1944. The diversity of vehicles is typical and interesting. Amongst them are a Citroën-Kégresse half-track, raid cars on 1936, 1937 and 1939 Chevrolet car chassis, an ex-French Army Studebaker truck (right), an *'Einheitsdiesel'* and a Panhard 178 armoured car (behind bus).

Part of a huge fleet of *Wehrmacht* vehicles turned over to the US Third Army after the surrender of the German Sixth Army in Austria. Picture was taken near Leizen on 12 May, 1945. Many of these vehicles were later released for civilian use in Austria.

INDEX

ACKNOWLEDGEMENTS

Colour drawings by C. A. J. van Woerkom, Olyslager Organisation (27).
Other illustrations: Messrs. John Chisman (1), Cyril Groombridge (2),
Yasuo Ohtsuka (2), Stanley C. Poole (1), Fernand Tiquet (1), and Laurie
A. Wright (1), Archivo General de la Nacion (2), Autocar (1), British
Official/ Imperial War Museum (8) and US Official (10). Remainder:
B. H. Vanderveen Collection. Thanks are also extended to Messrs. P.
Chamberlain, D. J. Voller and B. T. White for assistance rendered in the
compilation of this book.

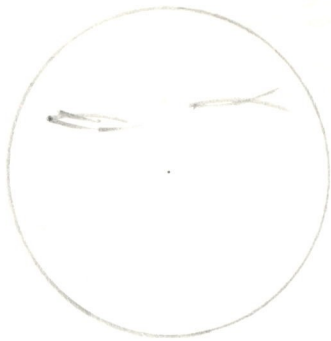